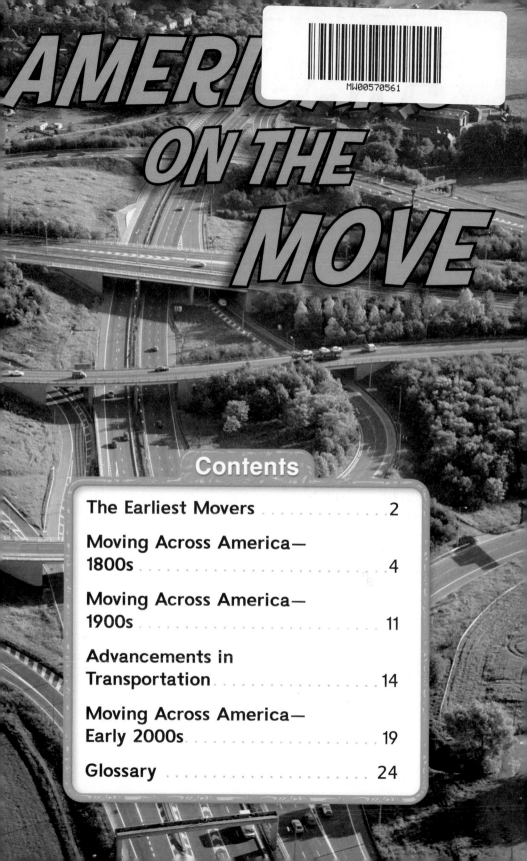

AMERICANS ON THE MOVE

Contents

Chapter 1

The Earliest Movers

People long ago walked from place to place. They used long pieces of trees to pull their goods. These tree pieces did not have wheels. But they were the first wagons. The Sumerians made the first wheels about 3000 B.C. Those wheels were made of stone. Then the Egyptians and Greeks made chariots. The Chinese built the first road system in about 1000 B.C.

The first four-wheeled carriages were used in the 12th century. Only rich people had carriages.

By the 18th century, more people were using transportation. Wagons and stagecoaches were two main **modes of transportation** used by the first American settlers in the mid-1800s.

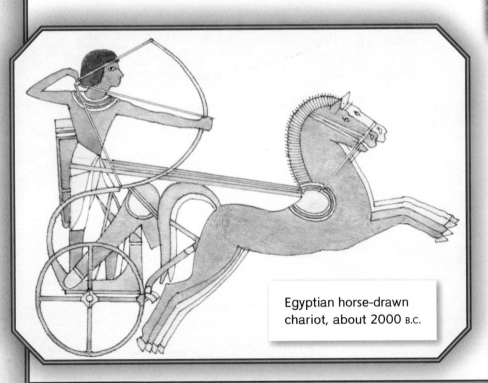

Egyptian horse-drawn chariot, about 2000 B.C.

Copyright © McGraw-Hill Education. (t to r) Dynamic Graphics/ Jupiterimages; (b) Dorling Kindersley/Getty Images

Stagecoaches were used for transportation before automobiles became popular.

First road systems built by Chinese

Mass use of four-wheeled carriages

American settlers use wagons and stagecoaches

Sumerians invent the wheel

First four-wheeled carriages invented

000 B.C.   1000 B.C.   12th century   18th century   mid-1800s

3

Chapter 2

Moving Across America— 1800s

Colonists first arrived in America in 1620. By the mid-1800s, the west was expanding. People of the Old West did not travel by car or airplane.

Settlers traveled across the west in stagecoaches and wagons. Most goods were shipped by boat. For this reason, many of America's first cities were built near water. Dirt roads were made to connect towns.

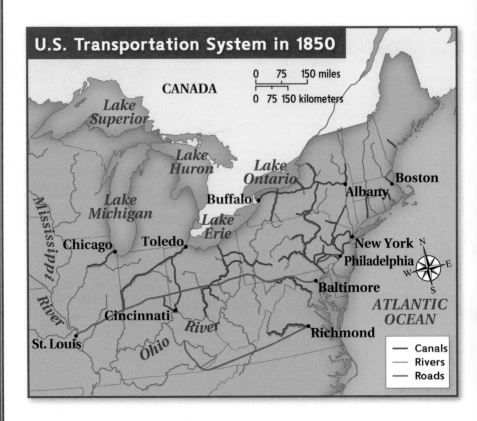

U.S. Transportation System in 1850

0   75   150 miles

0   75 150 kilometers

CANADA

Lake Superior

Lake Huron

Lake Ontario

Lake Michigan

Buffalo

Albany

Boston

Chicago

Toledo

Lake Erie

New York

Philadelphia

Mississippi River

Cincinnati

River

Baltimore

St. Louis

Ohio

Richmond

ATLANTIC OCEAN

N E W S

—— Canals
—— Rivers
—— Roads

People of the Old West often traveled on horseback.

Travel on dirt roads was difficult and slow. The railroad changed the lives of Americans. About 30,000 miles of track had been laid by 1860. Rail track construction started on both coasts. It met in Utah. On May 10, 1869, the world's first **transcontinental railroad** was completed.

Look at the bar graph. It shows how different kinds of transportation changed the lives of Americans between 1860 and 1870.

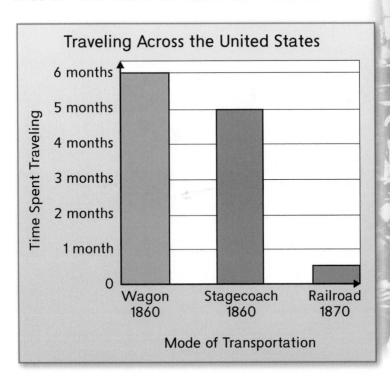

The final spike (the "Golden Spike") was struck on May 10, 1869, in Promontory Summit, UT. This railroad track connected the Mississippi River to the Pacific Ocean.

Railroads helped to ship goods faster. Fruits and vegetables grown in the west were taken by trains to the east. Goods made in the east were taken by trains to the west.

Travel by railroad was cheaper, too. Every year, more and more goods were shipped by train.

Tonnage of Freight on U.S. Railroads	
Year	Ton-Miles Shipped in Billions
1860	3.2
1870	9.0
1880	32.3
1890	79.1
1900	141.1

A *ton-mile* is one ton (2,000 pounds) carried one mile.

CURVE AT BROOKLY
NEW YORK & BROOKL

COPYRIGHT 1898
EO. P. HALL & SON.
PHOTOGRAPHERS.
NEW YORK

In the late 1800s, railroads connected American cities. People in cities rode streetcars. Before this, most people walked to work. They walked to shop. Now they could ride on the streetcar. People could work farther from where they lived. **Suburbs** resulted from this.

Streetcar

Did You Know?

In the late 1800s, bicycles were popular. Roads were usually rough for riders. So, bicyclists asked the government for better roads in the 1890s.

After a while, the bus and the car took the place of the streetcar. Buses and cars were faster and easier to use than streetcars.

A Quadricycle.

Chapter 3

Moving Across America— 1900s

Many different versions of the car developed in the late 1800s. Some early cars had three wheels. Early cars had different kinds of engines.

By 1894, the car looked more like the cars we see today. But it cost a lot of money to make and to buy cars. Most people could not afford a car. That changed in 1914.

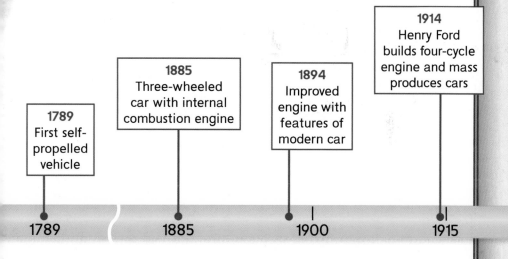

1789
First self-propelled vehicle

1885
Three-wheeled car with internal combustion engine

1894
Improved engine with features of modern car

1914
Henry Ford builds four-cycle engine and mass produces cars

1789   1885   1900   1915

In 1914, a man named Henry Ford changed how goods were made forever. Ford **mass produced** cars. He used **assembly lines** to make cars. Ford also gave people loans to buy cars. Middle-class people could now afford to buy cars.

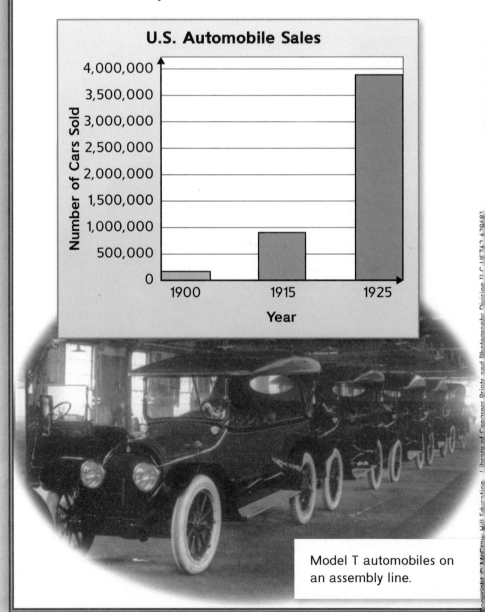

U.S. Automobile Sales

Model T automobiles on an assembly line.

Many Americans wanted to buy and drive cars. Cars let people travel freely. By 1929, Americans owned more than 5 million cars.

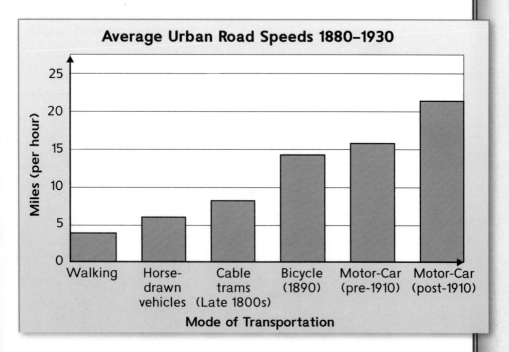

Average Urban Road Speeds 1880–1930

Miles (per hour) vs. Mode of Transportation

Mode of Transportation	Miles (per hour)
Walking	~4
Horse-drawn vehicles	~6
Cable trams (Late 1800s)	~8
Bicycle (1890)	~14
Motor-Car (pre-1910)	~16
Motor-Car (post-1910)	~21

Did You Know?

In the early 1900s, automobile advertisers marketed to children. Advertisers knew that children were the buyers of tomorrow. Toy cars were made soon after cars were on the market.

Chapter 4

Advancements in Transportation

Wilbur and Orville Wright made the first controlled flight in an airplane in 1903. The flight lasted only 12 seconds.

It took many years to make an airplane for **mass transportation**. By the 1930s, many Americans were able to travel by airplane.

The *Flyer* takes off with Orville Wright at the controls, as his brother Wilbur watches, December 17, 1903.

In the 1940s, the number of people who traveled by trains, airplanes, and cars rose. Transportation continued to improve.

This airplane is flying over a city in 1947.

Car travel got better as more roads were built. In 1956, the highway system began to be used. This system of roads connected American cities.

Interstate Highway System

CANADA

0   250   500 miles
0   250 500 kilometers

PACIFIC OCEAN

MEXICO

Gulf of Mexico

ATLANTIC OCEAN

— Interstate highway

41,000 miles of road connect cities.
Cost = $100 billion dollars

The **interstates** changed car travel in the United States. Interstate roads passed through small towns. Populations of these towns went down. Populations of the suburbs went up.

Rush hour on Interstate 405 in Los Angeles, CA.

Did You Know?

The most heavily traveled area of the interstate highway system is Interstate 405 in Seal Beach, Califonia. Approximately 377,000 automobiles travel this roadway each day. The least traveled area is Interstate 95 just north of Houlton, Maine. Approximately 1,880 automobiles travel this roadway each day.

In the 1960s, airplane travel began to change. The jet airplane was invented. The jet made air travel faster, cheaper, and more comfortable. More people began to travel by airplane.

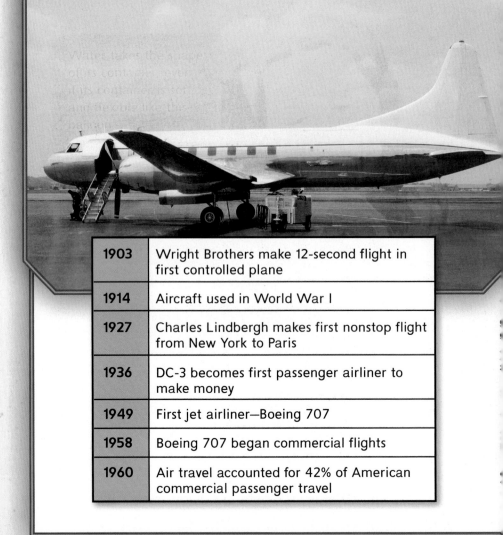

1903	Wright Brothers make 12-second flight in first controlled plane
1914	Aircraft used in World War I
1927	Charles Lindbergh makes first nonstop flight from New York to Paris
1936	DC-3 becomes first passenger airliner to make money
1949	First jet airliner—Boeing 707
1958	Boeing 707 began commercial flights
1960	Air travel accounted for 42% of American commercial passenger travel

Chapter 5

Moving Across America— Early 2000s

By the early 2000s, people traveled by train, car, and airplane.

Transportation in the Year 2000		
Automobiles	**Airplane Passengers**	**Railroad Passengers**
220 million cars owned	2 million passengers travel each day	21 million passengers travel each year

A busy day at Hartsfield International Airport in Atlanta, GA.

Americans are always traveling. Paying for transportation costs is hard. Roads and cars need to be safe. Railroads and airplanes also need to be safe.

Transportation Improvement Expenses	
Mode of Transportation	**Money Spent (millions)**
Air	$137,700
Highway	$1,336,200
Rail	$44,200
Water	$31,900

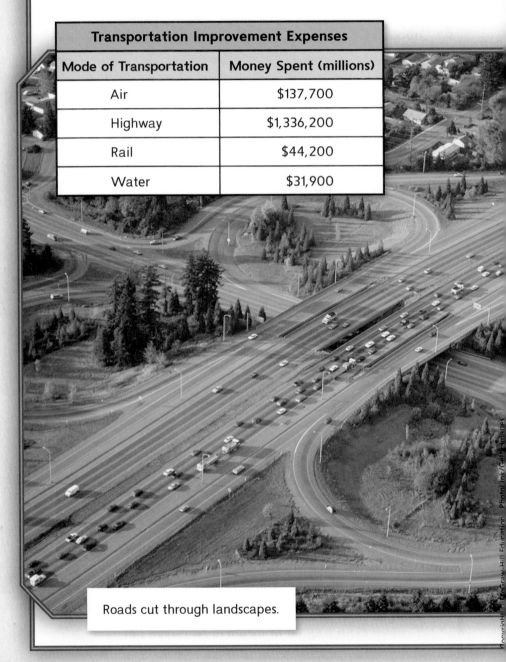

Roads cut through landscapes.

People are working to solve the problems caused by all these different modes of transportation. New standards are set for safer cars and roads. More people are riding buses again. This means less pollution and less traffic.

Mass transit refers to kinds of transportation, including buses, subways, and railroads.

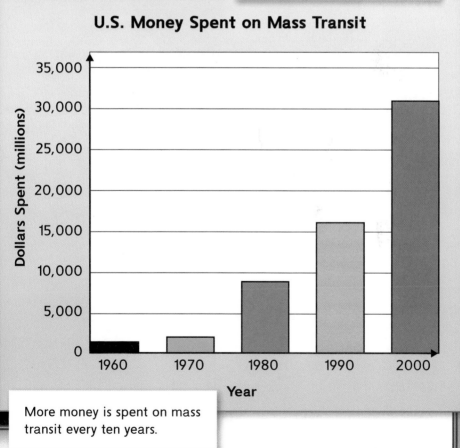

U.S. Money Spent on Mass Transit

More money is spent on mass transit every ten years.

What is ahead in the world of transportation? Time will tell.

Tourists test scooters in Washington, D.C.

Passengers board a bullet train.

The Moller M400 Skycar, a flying car, is able to drive at a speed of 326 miles per hour.

Glossary

assembly line
> A line of workers and machines along which a product moves as it is made. *(page 12)*

interstate
> A road that connects cities in two or more states with at least two lanes of traffic in each direction. *(page 17)*

mass produce
> The manufacture of large numbers of goods using identical parts and assembly-line methods. *(page 12)*

mass transportation
> A large number of goods or people moving from one place to another. *(page 14)*

mode of transportation
> A way of moving people or goods from one place to another. *(page 2)*

suburb
> A community outside of but near a larger city. *(page 9)*

transcontinental railroad
> A railroad that crosses an entire continent. *(page 6)*